# COLLOQUY ON MINORITY MALES IN SCIENCE, TECHNOLOGY, ENGINEERING, AND MATHEMATICS

Catherine Didion, Norman L. Fortenberry, and Elizabeth Cady, *Rapporteurs*

NATIONAL ACADEMY OF ENGINEERING
*OF THE NATIONAL ACADEMIES*

THE NATIONAL ACADEMIES PRESS
Washington, D.C.
www.nap.edu

**THE NATIONAL ACADEMIES PRESS 500 Fifth Street, NW Washington, D.C. 20001**

NOTICE: The project that is the subject of this report was approved by the Governing Board of the National Research Council, whose members are drawn from the councils of the National Academy of Sciences, the National Academy of Engineering, and the Institute of Medicine.

This project was supported by the National Science Foundation under award HRD-0533520. Any opinions, findings, conclusions, or recommendations expressed in this publication are those of the author(s) and do not necessarily reflect the views of the organizations or agencies that provided support for the project.

International Standard Book Number – 13: 978-0-309-26438-9
International Standard Book Number – 10: 0-309-26438-3

Additional copies of this report are available from the National Academies Press, 500 Fifth Street, NW, Keck 360, Washington, D.C. 20001; (800) 624-6242 or (202) 334-3313; http://www.nap.edu.



Printed in the United States of America

# THE NATIONAL ACADEMIES

*Advisers to the Nation on Science, Engineering, and Medicine*

The **National Academy of Sciences** is a private, nonprofit, self-perpetuating society of distinguished scholars engaged in scientific and engineering research, dedicated to the furtherance of science and technology and to their use for the general welfare. Upon the authority of the charter granted to it by the Congress in 1863, the Academy has a mandate that requires it to advise the federal government on scientific and technical matters. Dr. Ralph J. Cicerone is president of the National Academy of Sciences.

The **National Academy of Engineering** was established in 1964, under the charter of the National Academy of Sciences, as a parallel organization of outstanding engineers. It is autonomous in its administration and in the selection of its members, sharing with the National Academy of Sciences the responsibility for advising the federal government. The National Academy of Engineering also sponsors engineering programs aimed at meeting national needs, encourages education and research, and recognizes the superior achievements of engineers. Dr. Charles M. Vest is president of the National Academy of Engineering.

The **Institute of Medicine** was established in 1970 by the National Academy of Sciences to secure the services of eminent members of appropriate professions in the examination of policy matters pertaining to the health of the public. The Institute acts under the responsibility given to the National Academy of Sciences by its congressional charter to be an adviser to the federal government and, upon its own initiative, to identify issues of medical care, research, and education. Dr. Harvey V. Fineberg is president of the Institute of Medicine.

The **National Research Council** was organized by the National Academy of Sciences in 1916 to associate the broad community of science and technology with the Academy's purposes of furthering knowledge and advising the federal government. Functioning in accordance with general policies determined by the Academy, the Council has become the principal operating agency of both the National Academy of Sciences and the National Academy of Engineering in providing services to the government, the public, and the scientific and engineering communities. The Council is administered jointly by both Academies and the Institute of Medicine. Dr. Ralph J. Cicerone and Dr. Charles M. Vest are chair and vice chair, respectively, of the National Research Council.

**www.national-academies.org**

## NATIONAL ACADEMY OF ENGINEERING
## PROGRAM OFFICE STAFF

ELIZABETH CADY, Program Officer, National Academy of Engineering
CATHERINE DIDION, Senior Program Officer, National Academy of Engineering
CAMERON H. FLETCHER, Senior Editor, National Academy of Engineering
NORMAN L. FORTENBERRY[1], Director, Center for the Advancement of Scholarship on Engineering Education, Program Office, National Academy of Engineering (through April 29, 2011)
WEI JING, Research Associate, National Research Council
SIMIL RAGHAVAN, Associate Program Officer, National Academy of Engineering
PROCTOR REID, Director, Program Office, National Academy of Engineering
JASON WILLIAMS, Senior Financial Assistant, National Academy of Engineering

---

[1] Norman Fortenberry is currently the Executive Director of the American Society for Engineering Education.

# PREFACE AND ACKNOWLEDGMENTS

*Our two largest racial minority groups comprise about one third of the college-age kids in our country, and that fraction is steadily growing. But ... they earn less than 13 percent of the engineering degrees. Let me repeat this. The fastest growing segment of our young population earns less than 13 percent of our engineering degrees. Projecting forward, we have a workforce train wreck. We need to take action now to avoid it.*

– Charles M. Vest, President, National Academy of Engineering[1]

On August 8-12, 2010 the National Academy of Engineering (NAE), with funding from the National Science Foundation (NSF), convened the Colloquy on Minority Males in Science, Technology, Engineering, and Mathematics (STEM),[2] following the release of several reports highlighting the educational challenges facing minority males (e.g., by the College Board[3] and the National Center for Educational Statistics[4]). In addition, the American Institutes for Research (AIR) had noted in its 2007 report *Expanding the Pool of Potential STEM Graduates* that "underrepresented minority (i.e., African American, Hispanic American, and American Indian) males are leaking from the pipeline to STEM fields in higher education and beyond. The national trends are unmistakable."

Therefore NSF recognized the need to gather input from research communities that focus on minority males about how to frame investigations of gender-based factors that impact learning and choice in STEM education (both at the precollege and higher education levels) and the workforce for minority males. There was particular interest in framing a research agenda to study how interactions between minority males and societal and educational systems (both formal and informal) encourage or discourage the young men's interest and persistence in STEM. In addition, NSF hoped to gain community input to inform the parameters of a future NSF research program that could effectively address minority male participation in STEM. The Colloquy was held at the Mt. Washington Conference Center in Baltimore, Maryland, with approximately 40 participants, most of them researchers in education, psychology, sociology, mathematics, and physics. (The list of participants is in Appendix A.)

This report presents a summary of the Colloquy's breakout and plenary discussions, which addressed (a) research questions articulated in the breakout groups together with theories

---

[1] Charles M. Vest, "Engineers: The Next Generation - Do we need more? Who will they be? What will they do?" Speech, 2011 NAE Annual Meeting, October 16, 2011, Washington, DC.

[2] The Colloquy was supported by the National Science Foundation's Research on Gender in Science and Engineering (GSE) program and the Historically Black Colleges and Universities Undergraduate Program (HBCU-UP) of the NSF Directorate for Education and Human Resources (EHR) and the Social Psychology Program of the NSF Directorate for Social, Behavioral and Economic Sciences (HRD-0533520).

[3] The College Board, *The Educational Crisis Facing Young Men of Color*, January 2010. Available at http://professionals.collegeboard.com/profdownload/educational-crisis-facing-young-men-of-color.pdf.

[4] National Center for Educational Statistics, *Status and Trends in the Education of Racial and Ethnic Groups*, July 2010. Available at www.air.org/files/AIR-NCESracial_stats__trends1.pdf.

and methodologies to begin to address these questions; and (b) considerations for a potential research solicitation for the NSF, with major areas of inquiry concerning access, participation, and success for minority males in STEM.

This report reflects the views of the individuals who participated in the plenary and breakout groups. It has been reviewed in draft form by persons chosen for their diverse perspectives and expertise in accordance with procedures approved by the National Academies' Report Review Committee. The purpose of this independent review is to provide candid and critical comments that will assist the institution in making its published report as sound as possible and to ensure that the report meets institutional standards for quality and objectivity. The review comments and draft manuscript remain confidential to protect the integrity of the process.

We thank the following individuals for their review of the report: LeManuel Lee Bitsoi, Harvard University; Florence B. Bonner, Howard University; Daryl E. Chubin, American Association for the Advancement of Science; Lindsey Malcom-Piqueux, The George Washington University; Luis Ponjuan, Texas A&M University; and John Brooks Slaughter, University of Southern California. Although the reviewers listed provided many constructive comments and suggestions, they were not asked to endorse the content of the report, nor did they see the final draft before its release. The review of this report was overseen by Lance A. Davis, Executive Officer of the National Academy of Engineering, who was responsible for making certain that an independent examination of this report was carried out in accordance with institutional procedures and that all review comments were carefully considered. Responsibility for the final content of this report rests entirely with the authors and the institution. Any opinions, findings, and conclusions or recommendations expressed in this material are those of the authors and do not necessarily reflect the views of the National Science Foundation.

Catherine Didion
Norman L. Fortenberry
Elizabeth Cady, Rapporteurs

# CONTENTS

# 1

# INTRODUCTION

The National Academy of Engineering (NAE), with support from the National Science Foundation (NSF), hosted a Colloquy on Minority Males in Science, Technology, Engineering, and Mathematics (STEM) August 8–12, 2010, at the Mount Washington Conference Center in Baltimore, Maryland. The Colloquy was originally designed to frame a research agenda with respect to underrepresented minority males in science and engineering—African Americans, Hispanic Americans, and Native Americans. Discussions during the Colloquy resulted in the expansion of the populations of concern to include Native Pacific Islanders (Native Hawaiians and Polynesians) and Southeast Asian Americans (e.g., Filipino, Thai, and Vietnamese) as they too are often underrepresented in STEM fields.[1]

The Colloquy provided a forum for the identification of research theories and methodologies to help

- frame approaches to investigate race-, ethnicity-, and gender-based factors that impact learning and sustained interest in STEM education and the STEM workforce;
- encourage research examining differences within and among specific minority male populations; and
- enhance understanding of societal as well as formal and informal educational systems' interactions that encourage or discourage minority males' interest and perseverance in study or work in STEM fields.[2]

NAE staff reviewed recent research on minority males in STEM and also sought input from the NSF to draw up an invitation list with an eye toward balancing representation of research communities and minority male populations. Participants—primarily early career researchers in STEM, education, and the social and behavioral sciences—submitted information about how their work was relevant to the Colloquy's focus and how they hoped to leverage their attendance to further their research.

The first evening was an opportunity for the participants to meet each other and learn about their research (the agenda of the Colloquy is in Appendix B). The formal program was opened on the morning of August 9 by **Caesar Jackson**, Director of the Division of Human Resource Development in the NSF Directorate for Education and Human Resources (EHR). He

[1] The organizers of the Colloquy acknowledge that not all populations of minority males (e.g., Asian Pacific Islanders) were fully addressed at the Colloquy or, therefore, in this summary. This summary is intended as an initial step in future efforts to focus on engaging and encourage all populations to pursue STEM education and career paths.

[2] The focus of the Colloquy was on the broad framework of STEM education and careers. The discussions in the breakout groups emphasized STEM at the K–12 and the undergraduate levels.

welcomed attendees and emphasized the importance and relevance of the topic and the Colloquy to NSF's efforts in broadening participation in STEM education.

Next, **Jolene Jesse**, Director of NSF's Research on Gender in Science and Engineering (GSE) Program in the EHR Directorate, set the stage with a brief history of the GSE program. She observed that, as of the date of the Colloquy, her program at NSF which focuses on research on gender (i.e. on boys and men as well as on girls and women) had not received any proposals to address minority male participation in STEM, and that in her opinion while there had been progress on addressing girls' and women's participation in STEM, more research was needed on that of minority males. She presented the following goals for the Colloquy:

- To frame a research agenda on underrepresented minority males, addressing the following questions:
    - What do we know?
    - What do we need to know?
    - What would be key elements of an NSF solicitation to encourage research in this area?
    - What should be the ideal balance between research and implementation?
- To build bridges among researchers operating in different subspecialties of research on minority males with the aim of stimulating research collaborations and creating a sense of community.

# 2

# OPENING PLENARY

**Lorelle Espinosa**, Director of Policy and Strategic Initiatives at the Institute for Higher Education Policy in Washington, gave the plenary speech. She explained that more research on different populations will contribute to both the translation of research into practice and to the framing of theoretical work on the intersection of race and gender. Studies on men of color[1] can benefit, she said, from previous research and assessment of programmatic activities on gender that have focused on women and girls, adding that findings from prior theoretical work may be applicable to men and boys. For example, research on intersectionality[2] for minority males could refer to Black Feminist Theory as it looks at gender, race, and socioeconomic status. For research on men of color, Espinosa noted that it was important to take into account where they are in terms of their institution of higher education (e.g., community colleges and minority-serving institutions). It is also important to consider the geographic location of populations of young (precollege) minority males, as there are distinct geographic differences in the largest minority populations on the east and west coasts as well as the southern borders (see Figure 1).

---

[1] The terms "men of color," "minority males," and "underrepresented minority males" are used interchangeably in this report.

[2] The term intersectionality is used "to denote the various ways in which race and gender interact to shape the multiple dimensions" of an individual's experience. The term is often used in research on African American women. The core concept of intersectionality is that one cannot understand the full impact "wholly by looking at the race or gender dimensions of those experiences separately." Source: Kimberle Crenshaw, "Mapping the Margins: Intersectionality, Identity, Politics, and Violence against Women of Color," *Stanford Law Review*, Vol. 43, 1241, 1993.

**FIGURE 1.** Large metro areas with majority-minority child (under 18) populations, 2008.
SOURCE: Brookings Institution, 2010.

In response to an audience member's question, Espinosa agreed that it was important to look at research data on minority men in a disaggregated manner and that data for any single minority group may look very different from aggregated data across groups. Disaggregating the data by populations, regions, or ethnicities is critical to identifying target populations and capturing their unique characteristics as it relates to STEM participation. To illustrate, she referred to her slides on the top BS-granting colleges for minority males in STEM in 2007 (Table 1), showing that very few schools are listed as top producers for more than one minority male population.

**TABLE 1.** Top Ten BS-Granting Colleges/Universities for Minority Males in STEM, 2007

| | Black Men | Latino Men | Native American Men | Asian Pacific Islander Men |
|---|---|---|---|---|
| 1 | Southern University and A&M College | University of Puerto Rico, Mayaguez | Oklahoma State University | University of California, Berkeley |
| 2 | Florida A&M University | Florida International University | University of Oklahoma, Norman | University of California, Davis |
| 3 | North Carolina A&T State University | University of Texas, El Paso | Southeastern Oklahoma State University | University of California, San Diego |
| 4 | University of Phoenix | Universidad Politecnica de Puerto Rico | Northeastern State University | University of California, Irvine |
| 5 | Strayer University | University of Texas, Austin | North California State University, Raleigh | University of California, Los Angeles |
| 6 | Alabama A&M University | University of Florida | University of North Carolina, Pembroke | University of Texas, Austin |
| 7 | Howard University | University of Texas, Pan American | Arizona State University | University of Washington, Seattle |
| 8 | Prairie View A&M University | Texas A&M University | University of Arizona | San Jose State University |
| 9 | University of Florida | California State Polytechnic University, Pomona | East Central University | California State Polytechnic University, Pomona |
| 10 | University of Maryland, Baltimore County | The University of Texas, San Antonio | University of Washington, Seattle | University of Illinois, Urbana-Champaign |

SOURCE: Lorelle Espinosa, Institute for Higher Education Policy, Presentation on August 9, 2010.

# 3

# BREAKOUT SESSION 1: FOCUS ON RESEARCH POPULATIONS OF MINORITY MALES BY RACE AND ETHNICITY

The Colloquy included three breakout sessions with three or four discussion groups and rapporteurs responsible for summarizing and communicating the discussions in the plenary session. Each breakout group included senior researchers in the focal minority male population as topical resources. This report provides a summary of three breakout sessions that should give the reader a balanced picture of the conversations at the Colloquy.

The discussion groups in Breakout Session 1 considered the impacts of gender, ethnicity, and race in STEM education and research for minority populations. Breakout Session 2 focused on a discussion of theoretical frameworks. Participants in Breakout Session 3 considered the discussions of the previous sessions to identify research methodologies that could enhance understanding of minority male participation in STEM education and careers.

The groups in Breakout Session 1 were based on the racial and ethnic populations that were the primary focus of participants' research (e.g., African Americans, Native Americans and Pacific Islanders, and Hispanics). The breakout groups then considered the impact of gender on their selected population and its relevance and intersection with the ethnicity and racial identity of boys and men within these populations and discussed the following questions:

- **In what areas do gender differences exist?** What significant gaps exist in the research base with respect to the discovery and description of gender-based differences and preferences in learning STEM subjects at K–16 and in the graduate/faculty levels in the racial and ethnic population selected?
- **What is causing these gender differences?** What significant gaps exist in the research base with respect to understanding factors that affect interest, performance, and choice of STEM academic programs and careers in those fields where significant differences in participation and performance by gender exist in the selected minority male population?
- **How are gender differences exacerbated by educational settings?** What significant gaps exist in the research base with respect to discovering and understanding how experiences and interactions in informal and formal educational settings either inhibit or encourage interest and performance of learners based on gender within the selected minority male population?

In discussing these questions, most of the participants rejected deficit models[1] and agreed that the challenge was not to "fix" minority males but rather to create environments more conducive to their participation and performance in STEM. They identified the following ways to improve research activities: (a) refine studies of identity formation and consider a developmental perspective on identity (which would include gender, race, masculinity, and class); (b) ensure that current research findings are used to inform current practices to engage minority males in STEM and that the resulting challenges or failures of these practices are considered in future research; and (c) differentiate research questions and approaches between minority groups.[2]

**Breakout Group 1A: Researchers with a Focus on African American Males**

Participants noted that the curricular and cocurricular experiences of African American males may encourage or discourage their interest in STEM careers, and therefore underscored the importance of holistic approaches[3] for the recruitment, retention, and graduation of African American males in STEM fields at the undergraduate level.[4] Some participants in this group observed that efforts to help undergraduate African American males persevere and graduate are more effective when coupled with a knowledge and understanding of the precollege experiences of these young men. Accordingly they argued that research on the participation of African American males at graduate and professional levels in STEM should similarly focus on the factors that motivate their choice and perseverance in these education and career pathways.

Discussions were aligned with the educational levels (K–8, high school, undergraduate, and graduate/professional) identified by the Colloquy organizers. A summary of points raised for each level follows.

*For K–8:*

- To be more effective, different—and more qualitative—approaches are necessary for research on younger boys of color. Scale-based or other quantitative research studies may not be as helpful as qualitative methods, which should also consider age-appropriate developmental perspectives of identity.
- Key research objectives include identification of the leverage points for effective intervention to enhance academic performance and STEM interest. For example, peer connections may particularly merit investigation as one such point of leverage.

---

[1] According to the University of Kansas eLearning Design Lab, "The 'deficit' model focuses on the student as the major problem, neither looking within the environment nor the instructional practices in the classroom" (www.elearndesign.org).

[2] The following summaries of the breakout group discussions and suggested research topics are based on the presentations by the rapporteur of each group and should not be construed as consensus recommendations of the individual breakout groups, the Colloquy participants as a whole, or the National Academy of Engineering.

[3] According to Nandish Pantel in an article in the *International Journal of Education Management* 17 (6/7), "A holistic approach develops students to be critical, confident and independent. It aims to make learning a process of self-improvement that explicitly recognizes the self and the social context of learning and teaching, and recognizes the needs of the individual learner in the interaction."

[4] Undergraduate education in STEM discussed at the Colloquy includes both two- and four-year programs and the term "undergraduate" is used in this report to include both two-year and four-year schools.

- There is a need for more comparative and longitudinal studies on effective content and pedagogy, particularly those that promote earlier interest in STEM. Based on their current social prominence, two potential candidates for such studies include (a) STEM concepts present in communication technologies and social media (e.g., iPods, video games, and social media such as Facebook), and (b) the role and effectiveness of same-sex schools.

*For high school:*

- There are a number of questions that could focus potential research on this population, such as: What characterizes an educational culture of success? What pathways toward high school academic success are enabled or precluded by a child's elementary and middle school experiences, behaviors, and assessments? What are empowering, culturally relevant pedagogies that foster future STEM achievement? In what learning spaces (in and out of school) are they practiced?
- There is a lack of understanding by researchers of the interactions of racial identity and social capital theories.[5] Researchers must better understand how social and cultural capital develop and manifest in the academic and life trajectories of males of color and the multiple dimensions of identity among males of color at this level.

*Undergraduate (two- and four-year) education:*

- Research on students at this level is necessarily coupled with an understanding of precollege experiences that do or do not adequately prepare African American males for pursuing STEM study and careers. Such an understanding requires assessment of the nature and effectiveness of guidance in high school as well as the effectiveness of various school models (e.g., magnet schools, charter schools, and learning communities in conventional schools).
- There should be holistic approaches to understanding undergraduate recruitment, matriculation, retention, and graduation of African American males. In addition, there may need to be analysis of individual STEM disciplines in order to understand movement of African American males across these disciplines at the undergraduate level as well as their exit from STEM disciplines to non-STEM disciplines.

*For graduate students/professionals in STEM:*

- The future engagement of African American males in STEM fields is highly dependent on the presence of a racially diverse professoriate. Thus appropriate topics for research might include (a) the recruitment and retention of minority male faculty and (b) support structures to ease the transitions of African American males through these critical education and training milestones in STEM careers.

---

[5] Social capital theory is broadly defined as encompassing the "norms and networks facilitating collective action for mutual benefit." Source: Michael Woolcock, "Social Capital and Economic Development: Toward a Theoretical Synthesis and Policy Framework," *Theory and Society*, Vol. 27, 151-208, 1998.

**Breakout Group 1B: Researchers with a Focus on Hispanic American Males**

The participation and performance of Hispanic American males in STEM education and careers are imperiled by inequities that begin in elementary education systems and are reinforced by the culture and climate of academic and professional organizations, according to some participants in this breakout group. For example, educational settings exacerbate differences in gender performance (among all races) through differential expectations of teachers and faculty in gateway and gatekeeper courses.[6] In the Hispanic American community, these inequities are reinforced by differences in peer and parental influences (e.g., common cultural expectations that boys will leave educational pathways at the high school or baccalaureate level to assume gainful employment). During discussion, the group identified seven potential research strands to enhance understanding of these factors with respect to Hispanic American males:

- *K–8 academic preparation:* Hispanic American males enter the academic pipeline with high aspirations but too often leave with failure. Researchers need to understand why. Several factors may be at play: there appear to be challenges with the mathematics core (algebra, geometry, trigonometry, and precalculus); inadequate parental knowledge of academic requirements may be a contributory issue; and the language of instruction can be a challenge for second-language learners. Researchers also need to better understand the role of standardized testing and accelerated/advanced courses in systemic inequity and barriers to further STEM study by Hispanic males.
- *Successful navigation of undergraduate pathways:* Researchers should seek to better understand the messages students are getting in high school and as undergraduates about STEM study and careers. Studies should assess the impact of transfers from community colleges to four-year institutions, categorized by scientific field. And it would be useful to determine whether the NSF Research Experiences for Undergraduates Program[7] serves Hispanic males as well as it serves the general population.
- *Academic institutional policies and practices that (positively or negatively) affect Hispanic American males' attainment of undergraduate degrees in STEM:* It is important to identify models of institutions and programs that are effective at engaging Hispanic males at the undergraduate levels. How scalable are such programs? How might they be adapted, as appropriate, from ad hoc pilots to institutionalized programs? It is especially important to determine what policies and procedures encourage or inhibit faculty to support the recruitment and retention of graduate students of color, for example through mentoring and other supportive activities.
- *Understanding resiliency:* More information is needed about the characteristics of STEM achievers. What are their educational experiences? the characteristics of their home environments? Which ones build and nurture resiliency?
- *Identity formation:* There are numerous questions about identity formation among Hispanic American male students. How do they see themselves as scientists and engineers? Are such views congruent with their views of masculinity? What are the

[6] Gateway courses provide foundational experiences that can lead students to pursue further studies in STEM fields. Gatekeeper courses are required for advancement in STEM and often have a low rate of students passing.

[7] The Research Experiences for Undergraduates (REU) program is just one of many NSF-supported programs that are relevant to the discussions at the Colloquy. Another NSF program is the Alliances for Graduate Education (AGEP) Broadening Participation Research in STEM Education. Information is available at www.nsf.gov.

effects of interactions with teachers and/or coaches, and how are those interactions affected by the identity roles assumed by the students and their teachers and coaches? What is the influence of early immersion programs in mathematics or in science in the development of Hispanic males' sense of a "science identity"[8] and their success in STEM?

- *Effective pedagogy and instruction:* What are the components of models of success at the precollege and undergraduate levels? What is the role of culturally relevant pedagogy? How important are early immersion research opportunities in retaining Hispanic American males? What are particularly effective models of foundation mathematics and English instruction in the context of STEM?
- *Mentoring:* The most significant questions in this strand concern the preparation of effective mentors and their engagement to reach more students. How are teachers and faculty enabled to mentor Latino men? How might miscommunications resulting from differences be minimized between faculty and students? What incentives exist for teachers and faculty to serve as mentors and how effective are such incentives? What are the characteristics of effective mentors and how can they be replicated? How are teacher and faculty mentors best used without straining their professional, psychological, or physical well-being?

**Breakout Group 1C: Researchers with a Focus on Native American and Asian Pacific Islanders (Including Native Hawaiians)**

Some members of this breakout group noted that among Native Americans and Asian Pacific Islanders, societal and cultural issues underlie a variety of other significant challenges (e.g., low socioeconomic status and associated problems with community safety, school quality, and teacher quality) to STEM participation and performance. Research may need to focus on the following questions:

- What drives low rates of educational interest and attainment of Native American males as well as high rates of incarceration and military service? What are the language and cultural barriers? In particular, what steps can be taken to reverse the commonly held view that higher education, particularly for Native American males, is a selfish pursuit that does not contribute to the general welfare of their families and communities?
- What is the role of education in traditional-culture versus dominant-culture settings? How does the lack of household and educational resources (often meager in comparison to those of majority communities) affect Native American males' participation in STEM?
- How can pursuit of STEM education be facilitated for Native American males?

Participants in the breakout group articulated a number of challenges specific to Native American males. For example, unlike the other underrepresented minority populations, Native Americans have sovereign political identities and may identify as Native American, indigenous, or by tribal affiliation. Similarly, there are very distinct groups in Native American communities,

---

[8] Another important concept is self-efficacy, which is "belief in one's ability to perform a specific task." It can affect the goals one sets and "is related to the adoption of more challenging goals and greater commitment to those goals." See the *Self-Efficacy in STEM Information Sheet*, a product of the NAE and SWE-AWE. www.engr.psu.edu/awe/misc/ARPs/ARP_SelfEfficacy_InfoSheet_122208.pdf.

with unique challenges, that could benefit from research and interventions beyond those developed for African American and Hispanic American males. Furthermore, Native American males share a lack of visibility and are often misrepresented and least understood because of widespread lack of knowledge about the diversity of tribal nations in the United States. At the individual level, they have somewhat fluid definitions of who they are—how they self-identify—and this can impact how society views them.

Some participants did not believe it was appropriate for Native Americans and Pacific Islanders to be placed in a single breakout group and argued for distinctions among Native Americans, Native Pacific Islanders, and Asians to recognize their unique issues. Native Pacific Islanders do not have federal recognition of sovereignty. Asian Pacific Islanders are often not viewed as minorities and have struggled with recognition. The Asian population includes US-born and immigrant populations that are very diverse (e.g., Hmong) and are often lumped into the category of "Asian" without acknowledgment of how diverse these communities are.

## Recurring Themes

During the plenary session following the breakout groups, four common themes emerged in the rapporteurs' remarks:

- Research is needed to define the components of demonstrated models of success for minority males in STEM.
- Clear conceptualization of the challenges and positive factors that impact the academic success of minority males in STEM could result in powerful new models and theoretical frameworks.
- Research is needed to enhance understanding of the experiences of boys of color both within and across racial and ethnic groups, including self-identity and how it affects decision making about degree and career aspirations in STEM.
- Race and ethnicity are not always well defined and too often groups of underrepresented minority males are lumped together in categories that do not facilitate understanding.

# 4

# DAY 2 PLENARY

The Colloquy began the second full day with a talk by **James Stith**, former Vice President of the American Institute of Physics. Stith reflected on his career as a Black physicist who began as the child of a single mother with a third-grade education who was told that he would “never amount to anything.” He has taught physics at West Point and at the Ohio State University, and said that for his entire career he has been working on replacing himself. He believes that minority groups should be represented in the STEM professions proportional to their representation in the general population.

Stith explained that, since physics is usually the last science course students take in high school, he uses it as a bellwether for college readiness to enter STEM. In 1948 just over 25 percent of all US high school students had taken a physics course, in 2009 it was 36 percent (Figure 2).

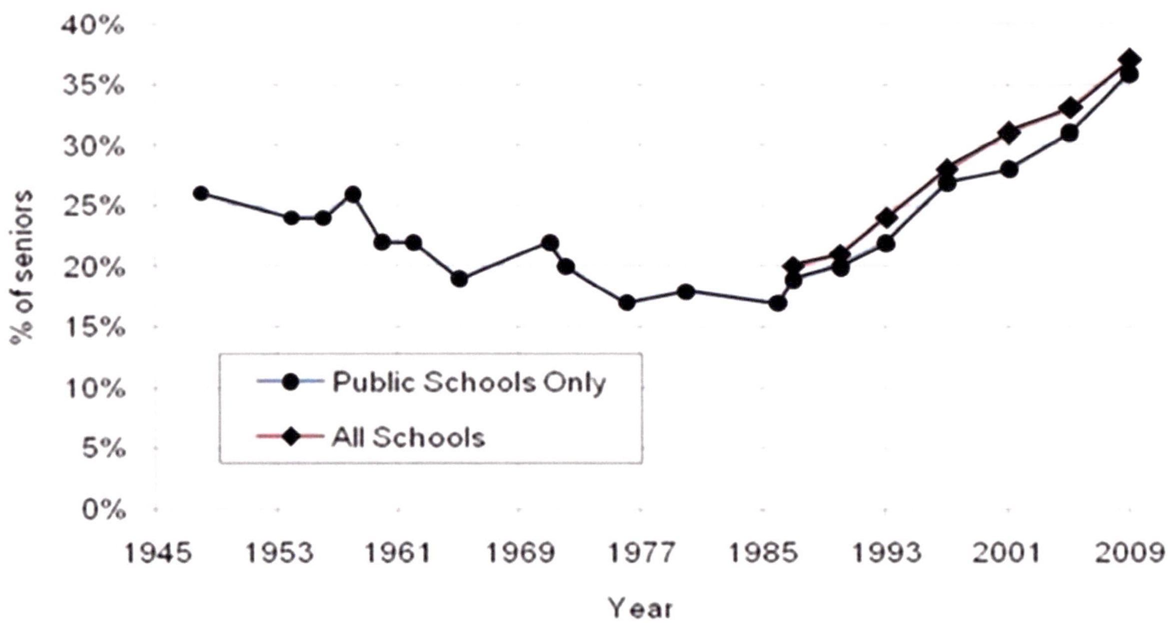


**FIGURE 2.** Physics enrollment in US high schools, 1948-2009. Data show percent of seniors who have taken at least one physics course prior to graduation.
SOURCE: 1987-current data were retrieved from American Institute of Physics (www.aip.org/statistics); data prior to 1987 were retrieved from National Center for Education Statistics (NCES; http://nces.ed.gov/).

However, he noted that, although the percentage and numbers of students who have taken a high school physics course have increased across all ethnicities and races over the past 25 years, there has not been similar growth for minorities who pursue physics at the undergraduate level: in 1996, roughly 6 percent (166 majors/year) of all physics baccalaureate recipients were African American; in 2007, it was less than 3 percent (144 majors/year) (Figure 3). The drop in percentage is due to both an increase in the total number of undergraduate physics degrees obtained by students of any race or ethnicity and a real decrease in the number of African American males attaining these degrees.

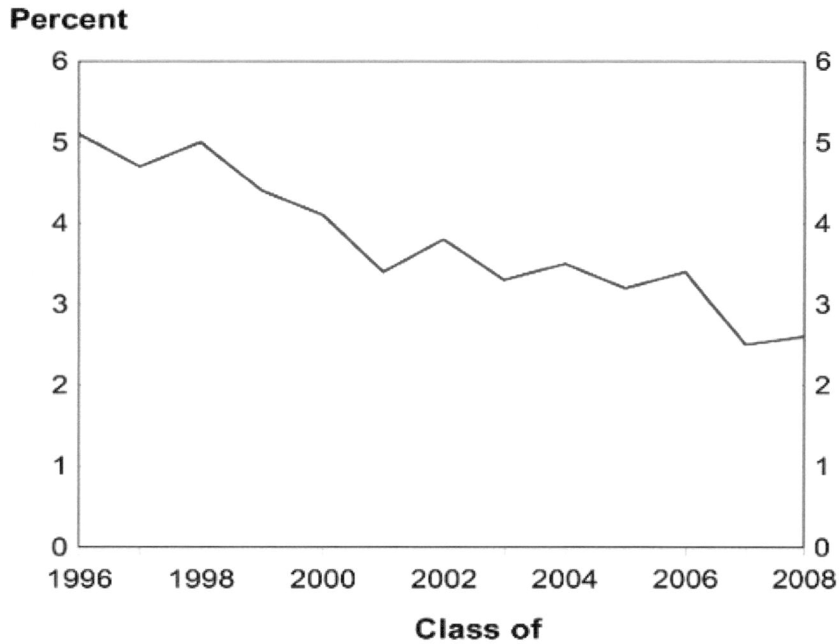


**FIGURE 3.** Percent of physics bachelor's (US citizens only) who were African American, classes 1996 through 2007.
SOURCE: American Institute of Physics (www.aip.org/statistics).

According to Stith, most African American physics PhD recipients used to receive their doctoral degrees from one of 34 historically black colleges or universities (HBCUs), which graduated roughly 60% of the African American baccalaureates in physics (Box 1). But because of program closures, only four HBCUs—Howard University, Hampton, Florida A&M, and Alabama A&M—now produce the majority of African American physics PhDs from all institutions. Stith found this change particularly troubling because only about 13% of African Americans currently attend any of the 110 HBCUs for their undergraduate education.

In conclusion, Stith stated that the continuous encouragement of minority students and their access to high-caliber teachers and faculty are two crucial factors needed for minority males to be successful in STEM studies and careers.

**Box 1: Universities That Awarded the Most Physics PhDs to African Americans since 1998**

Physics departments in these twelve universities awarded more than 65 percent of all physics PhDs degrees earned by African Americans since 1998.

- Alabama A&M University
- Cornell University*
- Florida A&M University
- Georgia Institute of Technology
- Hampton University
- Howard University
- Massachusetts Institute of Technology
- North Carolina State University
- Stanford University*
- University of Alabama at Birmingham
- University of California at San Diego
- University of Michigan*

* Denotes universities that have two departments that award physics PhDs. The universities on this list reported conferring three or more physics PhDs to African Americans between 1998 and 2007.

Source: American Institute of Physics Statistical Research Center, Enrollment & Degree.

# 5

# BREAKOUT SESSION 2
# DISCUSSIONS OF THEORETICAL FRAMEWORKS

The aim of the second day's breakout groups was to identify theoretical frameworks that might guide research to answer questions raised the day before. Attendees in the four randomly assigned groups were asked to

- identify challenges,
- characterize how they are manifested in the target populations,
- examine underlying mechanisms and remediation strategies, and
- provide models of innovative and successful approaches to overcoming the challenges.

Breakout group participants were invited to provide a graphical representation of their proposed frameworks.

**Breakout Group 2A**

This group developed a graphic of a circle with four quadrants corresponding to the major areas of inquiry and showing relevant theoretical frameworks (Figure 4). The group members identified a challenge—lack of culturally responsive faculty (Quadrant 1)—to illustrate application of the framework to the academic performance of minority males at the undergraduate level. The challenge is manifested by toxic school cultures that marginalize minority males (Quadrant 2). The underlying mechanisms of the challenge include social conformity and lack of incentives for change, and possible remediation strategies are faculty training and implementation of equity audits (Quadrant 3). Two models to address the challenge are replication of support structures shown to be effective at HBCUs and the use of culturally responsive activities (e.g., equity scorecards) to incentivize change (Quadrant 4).

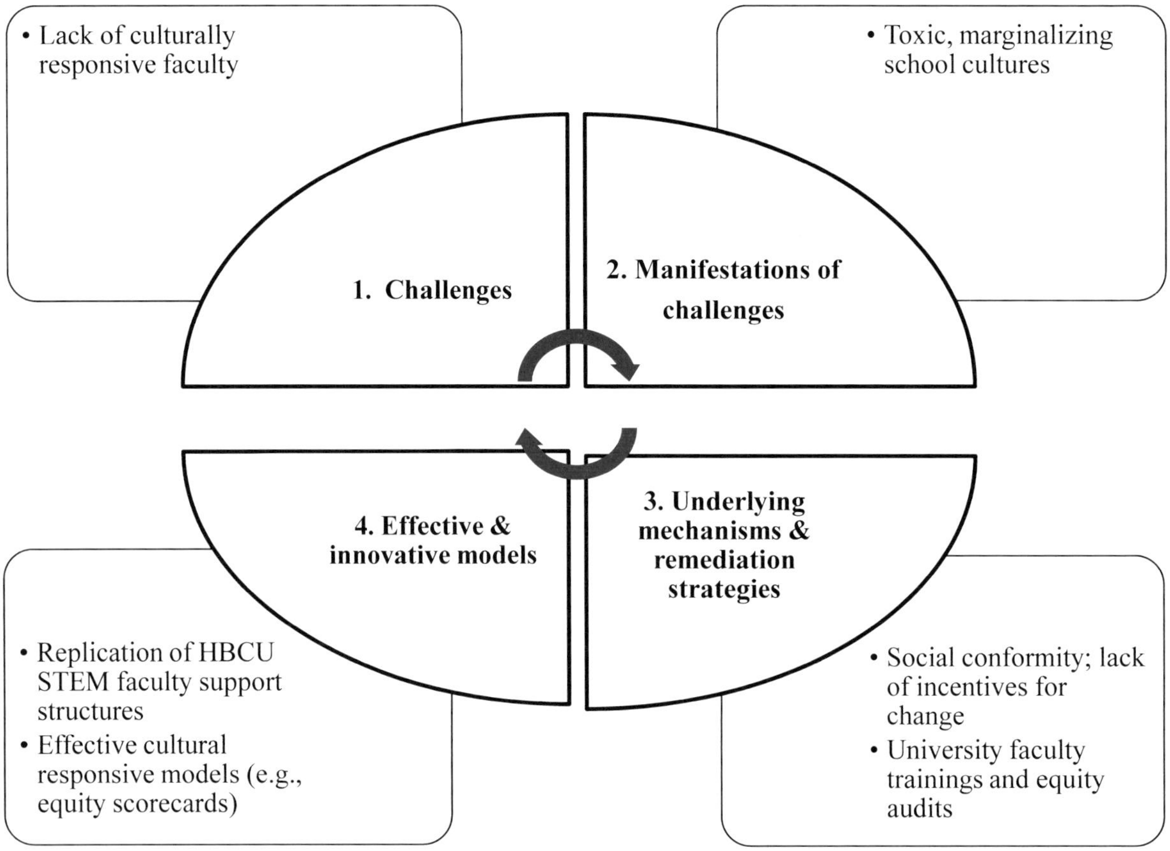


**FIGURE 4.** Breakout Group 2A's theoretical metaframework to address the challenge of having few culturally responsive faculty at the undergraduate level.

## Breakout Group 2B

This group developed a table with four rows corresponding to the major areas of inquiry, with accompanying relevant theoretical frameworks (Table 2) that are applicable for both precollege and postsecondary levels. For example, in the area of "Identify Challenges" the suggested frameworks are all in the broad category of Human Ecology, including attributional work (e.g., the work of Claude Steele) and Kurt Lewin's approach to social psychology.

**TABLE 2.** Metaframework Linking Major Areas of Inquiry with Theoretical Frameworks

<table>
<tr><td>Identify Challenges</td><td>Human Ecology:<br>• Attributional (encompasses Claude Steele's work)<br>• Margaret Beale-Spencer's PVEST Theory (1997)<br>• Urie Bronfenbrenner's Theory (1972)<br>• Kurt Lewin (social psychology)</td></tr>
<tr><td>Characterize Manifestations in Target Population</td><td>• Identify Theories<br>  - Race-based Identity Development theories including social construction of masculinity<br>  - Gender-based theories<br>  - Identity Development (Wortham)<br>    ▪ Adolescent Identity Development Theories<br>    ▪ Respectability<br>• Agency/Self-Efficacy/Self-Concept<br>  - Bandura, Mendosa, and Steele/Aaronson</td></tr>
<tr><td>Underlying Mechanisms (Models of Intervention)</td><td>• Discipline-based Intervention<br>  - Subject Matter Learning<br>• Organizational Theory<br>  - Organizational Psychology/Organizational Behavior<br>• Distributed intelligence with polled knowledge<br>• Social supportive models</td></tr>
<tr><td>Successful Examples</td><td>• Macro Examples:<br>  - The Algebra Project (Bob Moses)<br>    ▪ Organizational Psychology/Behavior Theories<br>  - The Meyerhoff Program (Hrabowski)<br>    ▪ Social Supportive Theories<br>  - DNIMAS (Norfolk State)<br>  - McNair Scholars / Summer Research Opportunities Program (SROP)<br>  - Others (e.g., program models for women in S&E)<br>• Micro (considered as concepts):<br>  - Models of Effort (Lauren Resnick)<br>    Models of Aspiration</td></tr>
</table>

SOURCE: Participants in Breakout Group 2B.

## Breakout Group 2C

Members of this group developed the graphic shown in Figure 5, which, unlike the others, is not tied to the four major research areas. Rather, it considers the individual in various contexts and interactions. The framework places particular emphasis on context as created by interactions among various metatheoretical frameworks related to an individual's social and cultural competency (e.g., one's identity, race and ethnicity, and social status).

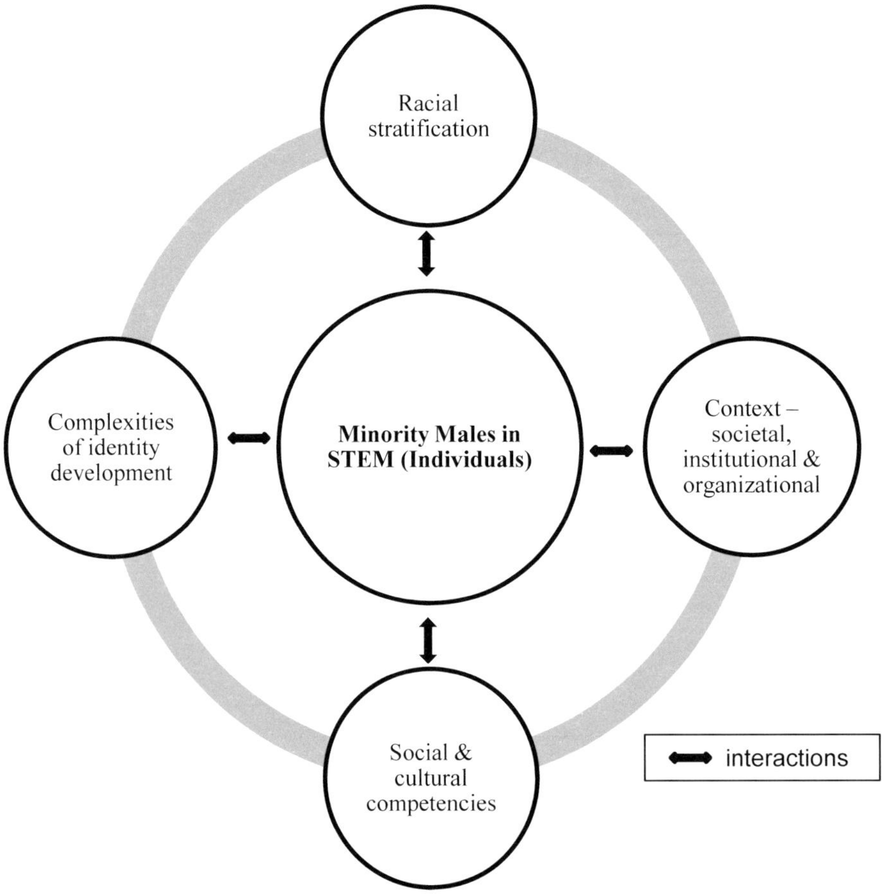


**FIGURE 5.** Theoretical framework connecting individuals, their interactions, and context. Developed by participants in Breakout Group 2C.

## Breakout Group 2D

This breakout group approached its task by developing a series of questions tied to the four major areas of inquiry (Box 2).

**BOX 2: Key Framework Questions for Each Major Area of Inquiry**

*Identify challenges:*

- Do male views of masculinity play into their decisions to pursue (or to not pursue) specific fields in STEM?
- What role do masculinity and gender play in the pursuit of science (e.g., the "feminization of science"; characterization of disciplines as "soft" vs. "hard")?
- What can be learned from a review of the data on students being "pushed out" or transferring from one science field to another? What role does microaggression play?
- How might deficit cognitive frame theory,[a] specifically with regards to faculty attitudes, improve understanding of the experience of minority males in STEM?

*Manifestations (characterize how challenges are manifested in target populations):*

- How might a review of cumulative advantage inform efforts to understand the experience of minority males? What are some best practices for creating a system of cumulative advantage in STEM for minority males? Are there different models at different educational levels?

*Mechanisms (examine underlying mechanisms and remediation strategies):*

- What layers of context should be taken into consideration in developing complex and comprehensive models of research interventions that include attention to individuals and families? How such models might be informed by a review of social, racial, policy, and ecological frameworks?

*Success models (provide models of innovative and successful approaches to overcoming the challenges):*

How might the following models (in whole or in part) improve understanding of what works to enhance the academic and career prospects of minority males?

- Resiliency and coping models
- Critical race theory (CRT), specifically with respect to interest convergence[b]
- "Academic identification," based on how well male students perform academically
- "Self theory," based on encouragement of students to see themselves in STEM programs and careers

[a]According to Estela Maria Bensimon ("Closing the Achievement Gap in Higher Education: An Organizational Learning Perspective" in *New Directions for Higher Education*, No. 131, Fall 2005, p. 103), deficit cognitive frame theory "focus[es] on stereotypical characteristics associated with the culture of disadvantage and poverty."
[b]According to interest convergence theory, substantive gains for minorities will occur only when they converge with the interests of the majority. See Derrick Bell's seminal article in the *Harvard Law Review* 518 (1980).

Source: Participants in Breakout Group 2D.

# 6

# BREAKOUT SESSION 3: DISCUSSIONS OF POTENTIAL RESEARCH METHODOLOGIES

After identifying theoretical frameworks, participants were randomly assigned to four groups of six to nine persons and turned their attention to identifying research methodologies for the questions raised the preceding day. Researchers in all four groups cautioned that the application of methodologies (and theories) must be driven by specific research questions and overall research design. Rapporteurs presented summaries of the groups' discussion in the afternoon plenary session, after which there was an opportunity for general discussion.

**Breakout Group 3A**

Members of this group articulated alternative ways of generating knowledge beyond traditional empirical research. One idea suggested was to combine and amplify quantitative and qualitative methods to get more information about the nuances of experience for men of color in STEM fields compared with many traditional empirical methods.

**Breakout Group 3B**

The discussion in this group focused on recognizing differences between standard research methodologies (e.g., structural equation modeling, hierarchical linear modeling, and classroom research) and "grassroots" methodologies (e.g., histories of scientific racism or bias). One idea suggested was to develop a new methodology incorporating both standard research and grassroots methodologies that could (a) be transformative and tied to action; (b) be collaborative, interdisciplinary, inclusive, and innovative; and (c) create new data sources.

**Breakout Group 3C**

Members of this group discussed the utility of a research design that supports examination of the individual, communal, organizational, and societal factors embedded in a major research area. Methodologies that would support such a broad research design would include many different types of research methods, including life event storytelling and surveying, longitudinal data collection, and ethnographies.

**Breakout Group 3D**

Participants in this group emphasized the need for new methodologies with appropriate theoretical grounding and pilot study validation. They emphasized the importance of broad diversity—in ethnicity, discipline, and background, as well as research theories and methodological experience—in the reviewer pool for any grant program, and suggested that experienced researchers should nominate other experienced researchers for service on review panels. Finally, they stressed that any research program solicitation should communicate that all research methodologies are equally valued and possibly provide examples.

# 7

# PLENARY: POSSIBLE NSF SOLICITATION SUPPORTING RESEARCH ON MINORITY MALES IN STEM

**Norman Fortenberry**, Director of the NAE Center for Advancement of Scholarship on Engineering Education (CASEE), led discussion of a possible NSF solicitation that might take into consideration the elements cited by many of the Colloquy participants. The discussion yielded the following observations:

- Contributions from a variety of research fields (including several in the social sciences) and a broad array of relevant theories, research designs, and methodologies could be of value and researchers from these fields should be encouraged to apply.
- Disaggregation of data could enhance the understanding of underlying dynamics that affect minority male participation in STEM studies, research, and careers.
- Innovative research designs that recognize and exploit small sample sizes inherent in the relevant communities might be useful.
- It may be helpful to give special attention to publicizing and providing guidance for the NSF solicitation to a broad community of researchers, with special efforts to engage diverse opinions and researchers who focus on underrepresented minority males.
- Research on the challenges and opportunities of practice could be highlighted as a priority in the solicitation.

Some attendees expressed concern about the adequacy of current NSF reviewer pools for proposals submitted to an NSF grant program addressing the research topics identified during the Colloquy, noting that it would be imperative for reviewers to be open to new approaches for research. In particular, given the small populations of many underrepresented minority males considered during the Colloquy, unconventional research methods might be required. An additional suggestion was that NSF sensitize its staff and reviewers to be receptive to unusual research approaches that may be unfamiliar or untraditional (compared with other NSF program solicitations) by providing information on the novel research methods and on the underrepresented minority male populations to be studied.

# 8

# ENLISTING RESEARCH SUPPORT FROM PRIVATE FOUNDATIONS

Attendees had opportunities during the Colloquy for informal discussions in self-selected groups that helped foster cross-disciplinary and cross-institutional research collaborations, which was one of the goals of the Colloquy stated by NSF Program Officer Jesse.

In support of these discussions **Robert Teranishi**, a consultant for the Ford Foundation's Advancing Higher Education Access and Success Initiative, made an impromptu presentation on grant opportunities available at the Ford Foundation, which, he indicated, is concerned about being responsive to needs of vulnerable populations. The foundation's agenda includes attention to college access and completion.

Teranishi urged resistance to attitudes such as "If you work at the margins, you'll get marginal change," which seem to sanction disregard for small minority communities. He strongly suggested both (a) engagement with program officers at private foundations because they often do not have external reviewers and (b) efforts to achieve a consensus discussion among foundation program officers, who may be more open to supporting work that uses the theoretical research frameworks discussed at the Colloquy.

# 9

# CLOSING PLENARY: LOOKING TO THE FUTURE – OTHER TOPICS TO CONSIDER

**Catherine Didion,** a Senior Program Officer of the NAE, introduced two overarching topics raised in plenary discussion sessions on the first day of the Colloquy that were tabled by general consensus until the closing plenary as many participants were concerned that they would require additional discussion time to adequately address.

- **Data collection and reporting.** Didion called attention to three specific challenges in this area: (a) For some minority populations, participation in STEM is sufficiently small that privacy concerns have prompted the practice of suppressing data on academic and professional progress. Yet such actions can significantly impede the ability to learn from promising and proven practices to increase participation in STEM. To inform and improve educational and professional practices, it is essential to achieve a balance between privacy concerns and researchers' access to valuable information. (b) Minority-serving institutions are significant sources of minority STEM baccalaureate recipients, yet their contributions are inadequately recognized and they are often not properly reflected in data collection and reporting. (c) Better data are needed for community college student populations. Data on community college contributions to STEM baccalaureate degree attainment are lacking even though some reports indicate that up to a third of community college students are students of color[1] and up to 50 percent of them aspire to transfer to a baccalaureate program. There is also concern about the definition of STEM. It clearly includes the physical and life sciences and engineering disciplines as well as technology fields associated with these disciplines, but does it also include management information systems or knowledge management systems? Analysts concerned about economic development and employment would expand the definition to incorporate many career and technical fields, whereas many traditional academics would not. In light of minority male interest in many technical fields and careers, should researchers view these fields as distinct, or as pathways to STEM, or as full partners of STEM? The utility of an in-depth discussion of the implications of such choices was suggested by several participants.
- **The role of for-profit institutions of higher education in the education of minority males.** Didion noted that data presented by Espinosa indicated that the online University of Phoenix and Strayer University were the fourth and fifth largest producers of STEM degrees for African American males in 2007 (Table 1). However, some for-profit educational institutions have engaged in practices that participants characterized as

[1] According to a 2012 report published by the Congressional Black Caucus Foundation, *Challenge the Status Quo: Academic Success Among School-Age African American Males* by Ivory A. Toldson and Chance W. Lewis, there are 1.2 million Black males in college and 43% attend a community college compared to 11% who attend a HBCU.

predatory. A future discussion should address (a) the relative merits of for-profit educational institutions in engaging and enrolling minority males, (b) public availability of the graduation rates of minority males from such institutions, and (c) the financial impact on minority males of choosing for-profit versus non-profit institutions for their education.

The following two additional concerns were raised during the closing plenary session:

- **Efforts are needed to ensure that research programs are tied to actions that demonstrably achieve positive outcomes for minority males.** There is a need for greater connectivity between research and implementation of programs based on the research.
- **Although much of the discussion at the Colloquy was framed in terms of boys and young men in formal or informal educational systems, many minority males have left or are in danger of leaving these systems,** suggesting the importance of continuous efforts to resolve surrounding issues that affect student enrollment, engagement, and completion of precollege education.

NSF Program Officer Jolene Jesse closed the Colloquy with an expression of appreciation for the participants' engagement in forthright discussions and their thoughtful deliberations. She indicated that she would explore the possibility of a distinguished lectureship series at NSF on this topic to better inform the NSF community about emerging research findings and possible opportunities through collaborative research ventures.

She encouraged attendees both to submit their names for consideration as potential reviewers for NSF grants and to explore possible collaborations with their fellow researchers at the Colloquy. Finally she noted that participants might investigate the possibility of NSF support for elements of their research identified during the Colloquy.

# 10

# EVALUATION OF THE COLLOQUY

*The Colloquy helped me to think theoretically about the challenges faced by different populations with respect to STEM. It also helped me to understand the ways in which the intersection of race, ethnicity, gender and other dimensions of identity impacts the experiences of men in science and engineering.*

– *Colloquy participant*

An evaluation[1] of the impact of the Colloquy, completed in 2012, noted that "respondents created a picture of a lively and collaborative environment growing from the conference," with 78 percent reporting that they "established or participated in new studies or exploratory collaborations" since attending the Colloquy and 74 percent reporting that they had developed new research or research collaborations as a result of the Colloquy. In addition, 55 percent of the respondents reported that they "changed their teaching practices as a result of exposure to STEM educational research."

The evaluation notes a large number of publications and presentations that resulted from the respondents' participation in the Colloquy, with submissions (39 percent) to funding agencies other than NSF and 21 percent to NSF. About a third (30 percent) of the respondents reported participating in NSF panels between the Colloquy and the evaluation (approximately 18 months).

The evaluator, Barbara Bogue noted in her report that "the responses of more than two-thirds of Colloquy participants two years after the event provide solid evidence that the Colloquy made a positive long-term impact on participant research, teaching, practice and career development."

The evaluation specifically reinforced the importance of networks in fostering collaboration among the researchers and in creating greater visibility for their research. And, based on the participants' comments about the impacts of the Colloquy on their research, the evaluation will be informative for future NAE projects that work toward both widening the talent pool of the engineering workforce and convening stakeholders to initiate actions to address diversity needs.

---

[1] For more information please see Barbara Bogue, *Post-Event Assessment Report: Colloquy on Minority Males in STEM, August 2010*, The AWE Project, Pennsylvania State University, July 2012.

# APPENDIX A

## LIST OF PARTICIPANTS

(Affiliation as of August 2010)

**Dorina Carter Andrew**
Michigan State University

**Lee Bitsoi**
Harvard University

**Jorge Chapa**
University of Illinois

**Michael Cunningham**
Tulane University

**James E. Davis**
Temple University

**Dedra Eatmon**
North Carolina A&T State University

**Lorenzo L. Esters**
Association of Public and Land-grant Universities

**Edmund W. Gordon**
Yale and Columbia Universities

**Shaun R. Harper**
University of Pennsylvania

**Paola Heincke**
The CEJJES Institute

**Lindsey E. Malcom**
University of California, Riverside

**Bryan T. Marks**
Morehouse College

**Daryl B. Holloman**
Columbus State University

**Jerlando F.L. Jackson**
University of Wisconsin, Madison

**Marlon C. James**
University of Connecticut

**Stanley L. Jackson, Jr.**
University of California, Los Angeles

**Amber Jones**
Emory University

**Tomashu "Kenyatta" Jones**
University of California, Los Angeles

**Tony Laing**
University of Illinois, Urbana-Champaign

**Nicole Lewis**
University of Kentucky

**Jonathan Lightfoot**
Hofstra University

**Arlene P. Macklin**
Morgan State University

**Tracy R. Rone**
Morgan State University

**Angela Slates**
University of Illinois, Urbana-Champaign

**Robert E. Megginson**
University of Michigan

**Ernest Morrell**
University of California, Los Angeles

**Christopher B. Newman**
University of California, Los Angeles

**Luis Ponjuan**
University of Florida

**Shameem Rakha**
University of Illinois, Urbana-Champaign

**David Wall Rice**
Morehouse College

**Dana Taylor**
North Carolina A&T State University

**Edd Taylor**
Northwestern University

**Robert Teranishi**
New York University

**William T. Trent**
University of Illinois, Urbana-Champaign

**J. Luke Wood**
Lincoln Memorial University

**Speakers:**

**Lorelle Espinosa**, Institute for Higher Education Policy
**James Stith**, American Institute of Physics (Emeritus)

**National Science Foundation Staff:**

**Kellina Craig-Henderson**, Program Director, Social Psychology Program, Social, Behavioral and Economic Sciences Directorate
**Caesar Jackson**, Acting Director, Division of Human Resource Development, Education and Human Resources Directorate
**Jolene Kay Jesse**, Program Director, Research on Gender in Science and Engineering, Education and Human Resources Directorate
**Claudia Rankins**, Program Director, Historically Black Colleges and Universities Program, Education and Human Resources Directorate

**National Academy of Engineering Staff**

**Norman L. Fortenberry**, Director, Center for the Advancement of Scholarship on Engineering Education
**Catherine J. Didion**, Senior Program Officer
**Elizabeth T. Cady**, Associate Program Officer
**Wei Jing**, Apprentice
**Jason Williams**, Senior Financial Assistant

# APPENDIX B

# COLLOQUY AGENDA

**Mt. Washington Conference Center**
**Baltimore, MD**
**August 8–12, 2010**

| **Sunday, August 8, 2010** | | |
|---|---|---|
| 6:00 – 7:00 p.m. | Introductions & Ice Breaker<br>Catherine Didion, *National Academy of Engineering (NAE)* | Conference Room 18 |
| 7:00 – 9:00 p.m. | Buffet Dinner with Cash Bar | Octagon Atrium |

| **Monday, August 9, 2010: Major Research Strands, Gaps, and Opportunities** | | |
|---|---|---|
| 7:00 – 8:00 a.m. | Breakfast served in McAuley Café | 1st Level, McAuley Hall |
| 8:00 – 8:20 a.m. | Welcome and Greetings<br>Norman Fortenberry, *NAE*<br>Jolene Jesse, *National Science Foundation (NSF)*<br>Caesar Jackson, Director, *Division of Human Resource Development, NSF* | Conference Room 19 |
| 8:20 – 8:30 | Goals for the Colloquy and Charge to Attendees<br>Jolene Jesse, *NSF* | Conference Room 19 |
| 8:30 – 9:00 a.m. | Plenary Speaker<br>Lorelle Espinosa, *Institute for Higher Education Policy* | Conference Room 19 |
| 9:00 – 10:00 a.m. | Breakout 1: *Focus on Ethnicity*<br>What are the gaps in our knowledge about<br>• The areas in which gender differences exist?<br>• What is causing gender differences?<br>• How gender differences are exacerbated by educational settings? | Individual Rooms<br>*(Breakout Groups and Rooms will be Assigned Monday morning – see handout for list of rooms)* |
| 10:00 – 10:30 a.m. | Break | |
| 10:30 – 11:30 a.m. | Breakout 2: *Focus on Educational Level*<br>What are gaps in knowledge with respect to particular questions on performance by gender and ethnicity that are unique to particular educational levels? | Individual Rooms<br>*(see handout)* |
| 11:30 – 12:30 p.m. | Breakout 3: *Synthesis by Ethnicity across Educational Level* | Individual Rooms |

| | | |
|---|---|---|
| | Return to Breakout 1 groups and update your findings to reflect key points from Breakout 2 discussions | *(see handout)* |
| 12:30 – 2:00 p.m. | Lunch | Hayward Dining Room<br>2nd level McAuley Hall |
| 2:00 – 2:30 p.m. | Report out from Breakout 3 | Conference Room 19 |
| 2:30 – 3:45 p.m. | General Plenary Discussion on key points from Breakout 3 | Conference Room 19 |
| 3:45 – 4:15 p.m. | Break | |
| 4:15 – 5:00 p.m. | Implications of Today's Discussion for Tuesday's Discussions | Conference Room 19 |
| 5:00 – 6:00 p.m. | Break | |
| 6:00 – 8:00 p.m. | Dinner | Hayward Dining Room |

**Tuesday, August 10, 2010: Theoretical Frameworks and Research Methodologies**

| | | |
|---|---|---|
| 7:00 – 8:00 a.m. | Breakfast served in McAuley Café | McAuley Hall |
| 8:00 – 8:30 a.m. | Summary of Day 1 | Conference Room 19 |
| 8:30 – 9:00 a.m. | Plenary Speaker | Conference Room 19 |
| | James Stith, *American Institute of Physics - Emeritus* | |
| 9:00 – 10:30 a.m. | Breakout 4: *Theoretical Frameworks* | Individual Rooms |
| | • What theoretical frameworks may aid in closing the gaps identified in Breakout 3? | *(see handout)* |
| | • Should new frameworks be synthesized or created? | |
| 10:30 – 11:00 a.m. | Break | |
| 11:00 – 12:00 a.m. | Breakout 5: *Research Methodologies* | Individual Rooms |
| | • What research methodologies may aid in explicating the theoretical frameworks identified in Breakout 4? | *(see handout)* |
| | • Should new methodologies be synthesized or created? | |
| 12:00 – 2:00 p.m. | Lunch | Hayward Dining Room |
| 2:00 – 2:45 p.m. | Report out from Breakouts 4 and 5 | Conference Room 19 |
| 2:45 – 3:45 p.m. | Plenary Discussion of Key Theoretical Issues Informed by Methodology Discussion | Conference Room 19 |
| 3:45 – 4:15 p.m. | Break | |
| 4:15 – 5:00 p.m. | Anticipating the implications of today's discussion for Wednesday's discussions | Conference Room 19 |
| 5:00 – 6:00 p.m. | Break | |
| 6:00 – 8:00 p.m. | Dinner | Mt. Washington Tavern<br>5700 Newbury Street |

**Wednesday, August 11, 2010: Framing a Research Agenda**

| | | |
|---|---|---|
| 7:00 – 8:00 a.m. | Breakfast served in McAuley Café | 1st Level, McAuley Hall |
| 8:00 – 8:30 a.m. | Summary of Tuesday<br>Goal of Symposium : *New Solicitation* | Conference Room 19 |

| | | |
|---|---|---|
| 8:30 – 10:00 a.m. | Discussion of Big Questions<br>What major questions should be reflected in a solicitation? | Conference Room 19 |
| 10:00 – 10:30 a.m. | Break | |
| 10:30 – 12:00 p.m. | Breakout 6: *Organizational Models and Change Process by Educational Level*<br>What organizational and change models may most aid in applying research findings to achieve improved practice? | Individual Rooms |
| 12:00 – 12:30 p.m. | Report out from Breakout 6 | Conference Room 19 |
| 12:30 – 2:00 p.m. | Lunch | Hayward Dining Room |
| 2:00 – 3:00 p.m. | Breakout 7: *Types of Project Teams*<br>What types of project teams should be encouraged/allowed? | Conference Room 19 |
| 3:00 – 3:30 p.m. | Report out from Breakout 7 | Conference Room 19 |
| 3:30 – 4:00 p.m. | Break | |
| 4:00 – 5:00 p.m. | Discussion – *What's Missing?* | Conference Room 19 |
| 5:00 – 6:00 p.m. | Break | |
| 6:00 – 8:00 p.m. | Dinner – Outdoor Barbeque | Conference Center Grounds |

## Thursday, August 12, 2010: Wrapping Up

| | | |
|---|---|---|
| 7:00 – 8:00 a.m. | Breakfast served in McAuley Café | 1st Level, McAuley Hall |
| 8:00 – 8:30 a.m. | Summary of Wednesday: Accomplishments | Conference Room 19 |
| 8:30 – 10:15a.m. | Summarizing Discussion, Observations | Conference Room 19 |
| 10:15 – 10:30 | Closing Comments<br>Jolene Jesse, *NSF*<br>Norman Fortenberry, *NAE* | Conference Room 19 |
| 10:30 a.m. | Adjourn | |
| 11:00 a.m. | Departure | |